MATHS Workbook

Level **1**

MOONSTONE

Published in Moonstone
by Rupa Publications India Pvt. Ltd 2022
7/16, Ansari Road, Daryaganj
New Delhi 110002

Sales centres:
Allahabad Bengaluru Chennai
Hyderabad Jaipur Kathmandu
Kolkata Mumbai

ISBN: 978-93-5520-724-1

First impression 2022

10 9 8 7 6 5 4 3 2 1

The moral right of the authors has been asserted.

Contents

Count the Objects

1. **Look at the picture and count the different food items which the baker is selling.**

2. **Write down how many of each type of food items are there at the bakery.**

(a)  ☐

(b)  ☐

(c) 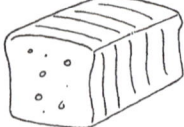 ☐

(d)  ☐

(e)  ☐

(f) 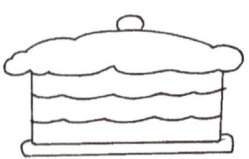 ☐

3. **Write the missing numbers in the number strip given below.**

(a)

| 1 | 2 | ___ | ___ | 5 | ___ | 7 | ___ |

(b)

| 7 | 8 | ___ | ___ | ___ | 12 | ___ | 14 |

(c)

| 13 | ___ | 15 | ___ | 17 | ___ | ___ | 20 |

(d)

| 0 | ___ | ___ | 3 | ___ | ___ | 6 | ___ |

(e)

| 7 | ___ | ___ | 10 | ___ | ___ | 13 | ___ |

Count and Match

1. Count the objects below and match each to the correct number and number names.

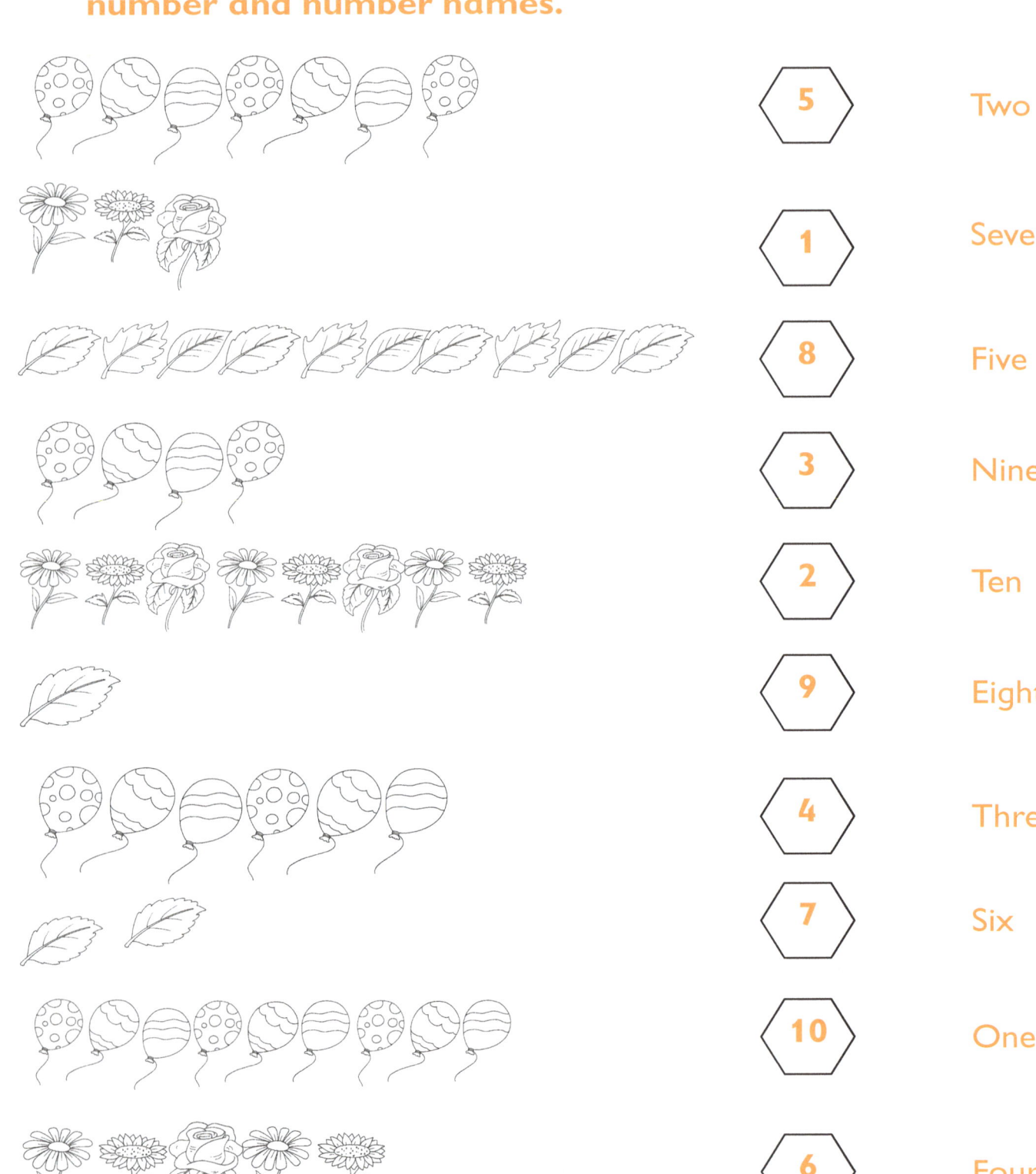

Forward and Backward

1. **Count forwards and write the missing numbers.**

(a)
0	1				5				9

(b)
4				8				12	13

(c)
10			13				17		19

(d)
20					25		27		29

2. **Count backwards and write the missing numbers.**

(a)
10				6			3		1

(b)
15			12			9			6

(c)
18				14			11		9

(d)
29		27		25					20

Before, After and Between Numbers

1. Write the number that comes before the given number.

(a)

(b)

(c)

(d)

(e)

(f)

(g)

(h)

(i)

(j)

2. Write the number that comes after the given number.

(a) 5 __
(b) 7 __
(c) 18 __

(d) 9 __
(e) 1 __
(f) 11 __

(g) 6 __
(h) 3 __
(i) 19 __

(j) 14 __
(k) 10 __
(l) 12 __

3. Write the number that comes between the given numbers.

(a) 15 __ 17
(b) 3 __ 5

(c) 8 __ 10
(d) 12 __ 14

(e) 9 __ 11
(f) 6 __ 8

(g) 1 __ 3
(h) 14 __ 16

Comparing Numbers

1. Pair each flower from group A to a flower in group B. Are any flowers left unpaired in either of the groups? Circle the group with unpaired flowers.

Set I

Set II

2. Compare the numbers given below and use the correct sign (> , < , =) in the ☐ .

(a) 7 ☐ 6

(b) 15 ☐ 15

(c) 6 ☐ 6

(d) 4 ☐ 2

(e) 8 ☐ 3

(f) 1 ☐ 5

(g) 9 ☐ 13

(h) 2 ☐ 11

(i) 18 ☐ 9

(j) 10 ☐ 10

Ordering Numbers

1. **Write the numbers in increasing order.**

(a)

7 3 5 2 ⟶ ◯ ◯ ◯ ◯

(b)

9 1 4 6 ⟶ ◯ ◯ ◯ ◯

(c)

1 8 10 3 ⟶ ◯ ◯ ◯ ◯

(d)

4 9 7 2 ⟶ ◯ ◯ ◯ ◯

2. **Write the numbers in decreasing order.**

(a)

8 5 0 4 ⟶ ◯ ◯ ◯ ◯

(b)

2 10 6 3 ⟶ ◯ ◯ ◯ ◯

(c)

7 4 9 1 ⟶ ◯ ◯ ◯ ◯

(d)

3 8 5 10 ⟶ ◯ ◯ ◯ ◯

Addition up to 10

1. **Count and add the objects to write the sum in the** .

(a) 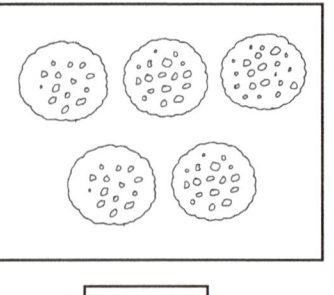 and 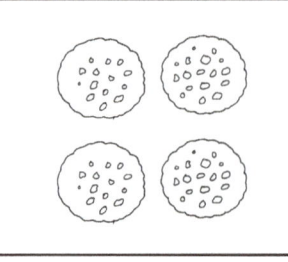 equals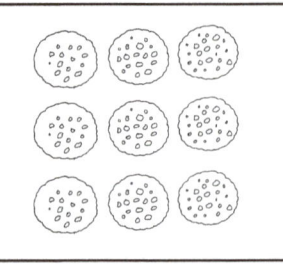

$\boxed{}$ **+** $\boxed{}$ **=** $\boxed{}$

(b) 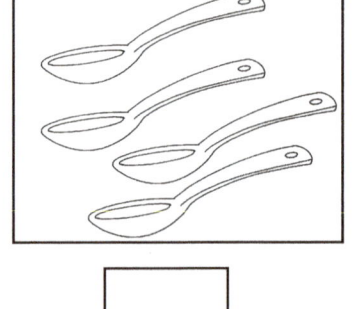 and 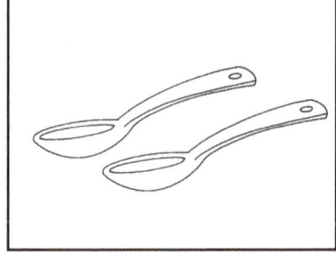 equals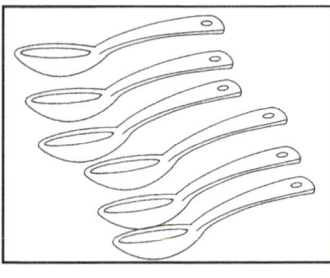

$\boxed{}$ **+** $\boxed{}$ **=** $\boxed{}$

(c) 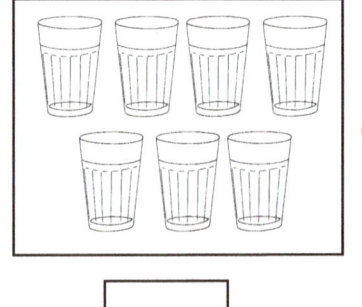 and  equals

$\boxed{}$ **+** $\boxed{}$ **=** $\boxed{}$

2. Add the numbers in each row and write their sum in each box.

(a) $6 + 1 = \boxed{}$ (b) $9 + 1 = \boxed{}$ (c) $4 + 0 = \boxed{}$

(d) $3 + 2 = \boxed{}$ (e) $5 + 3 = \boxed{}$ (f) $8 + 2 = \boxed{}$

(g) $1 + 8 = \boxed{}$ (h) $7 + 1 = \boxed{}$ (i) $3 + 3 = \boxed{}$

Add on the Number Line

1. Use the number line to add the numbers below and write their sum in the ☐.

(a) 4 + 3 = ☐

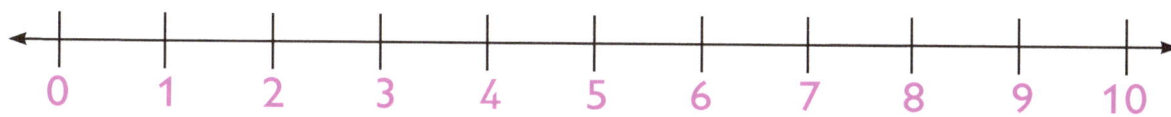

(b) 2 + 2 = ☐

(c) 8 + 1 = ☐

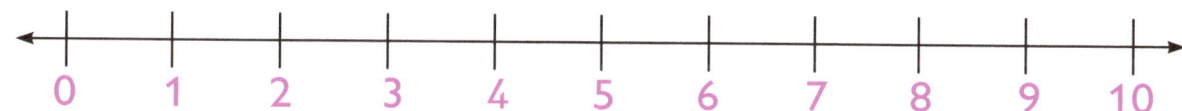

(d) 6 + 4 = ☐

(e) 5 + 1 = ☐

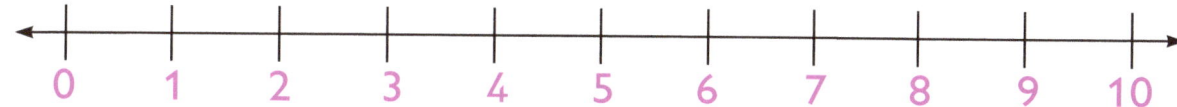

(f) 7 + 3 = ☐

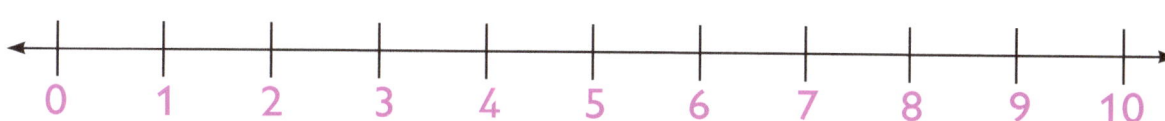

Subtraction up to 10

1. **Cross out according to the sum, then count the remaining numbers and write them in the ▢.**

(a)

From 8 pencils, take away 3.

$8 - 3 = \boxed{}$

pencils are left.

(b)

From 6 flowers, take away 1.

$6 - 1 = \boxed{}$

flowers are left.

(c)

From 9 apples, take away 5.

$9 - 5 = \boxed{}$

apples are left.

(d)

From 5 books, take away 2.

$5 - 2 = \boxed{}$

books are left.

(e)

From 7 pencils, take away 3.

$7 - 3 = \boxed{}$

pencils are left.

(f)

From 8 apples, take away 2.

$8 - 2 = \boxed{}$

apples are left.

(g)

From 9 books, take away 1.

$9 - 1 = \boxed{}$

books are left.

2. Subtract 1 from each number and write the answers in the ▢ .

(a) $9 - 1 =$ ▢   (b) $8 - 1 =$ ▢   (c) $2 - 1 =$ ▢

(d) $7 - 1 =$ ▢   (e) $3 - 1 =$ ▢   (f) $5 - 1 =$ ▢

(g) $4 - 1 =$ ▢   (h) $10 - 1 =$ ▢   (i) $6 - 1 =$ ▢

3. Subtract 0 from each number and write the answers in the ▢ .

(a) $8 - 0 =$ ▢   (b) $5 - 0 =$ ▢   (c) $1 - 0 =$ ▢

(d) $3 - 0 =$ ▢   (e) $9 - 0 =$ ▢   (f) $7 - 0 =$ ▢

4. Subtract the given numbers and write the answers in the ▢ .

(a) $9 - 4 =$ ▢   (b) $5 - 2 =$ ▢   (c) $8 - 1 =$ ▢

(d) $10 - 3 =$ ▢   (e) $6 - 4 =$ ▢   (f) $4 - 2 =$ ▢

(g) $7 - 1 =$ ▢   (h) $2 - 2 =$ ▢   (i) $3 - 3 =$ ▢

(j) $3 - 2 =$ ▢   (k) $8 - 6 =$ ▢   (l) $7 - 4 =$ ▢

Subtract on the Number Line

1. **Use the number line to subtract and write the answers in the** ⬜ **.**

(a) 7 − 2 = ⬜

(b) 4 − 4 = ⬜

(c) 8 − 6 = ⬜

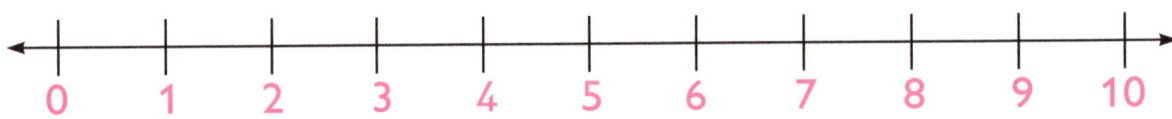

(d) 6 − 4 = ⬜

(e) 5 − 2 = ⬜

(f) 9 − 3 = ⬜

Addition Facts

1. The table below shows the addition facts for numbers up to 10. Complete the table by adding the numbers in rows and columns.

	1 ↑	2 ↑	3 ↑	4 ↑	5 ↑	6 ↑	7 ↑	8 ↑	9 ↑
9 →	10								
8 →									
7 →	8	9	10						
6 →			9						
5 →		7			10				
4 →	5			8					
3 →	4	5	6		8	9			
2 →	3	4	5				9		
1 →	2	3	4	5		7			10
+	1	2	3	4	5	6	7	8	9

2. Write the missing number in the addition statements given below.

(a) 6 + _____ = 10

(b) 3 + _____ = 10

(c) 5 + _____ = 6

(d) _____ + 2 = 6

(e) _____ + 4 = 7

(f) 6 + _____ = 7

(g) 2 + _____ = 4

(h) 4 + _____ = 4

(i) 3 + _____ = 9

Number Bonds

1. Write the missing numbers to complete each number bond.

(a)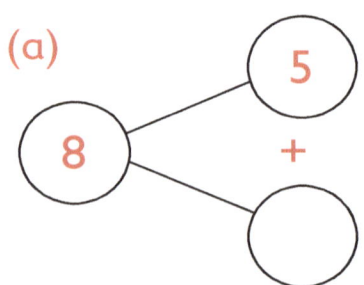

(b)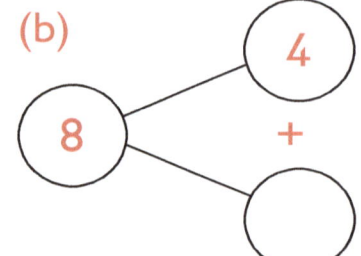

(c)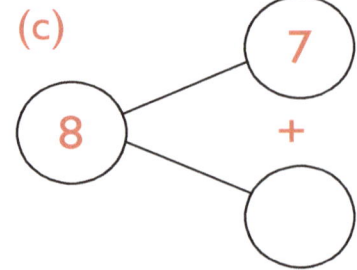

(d)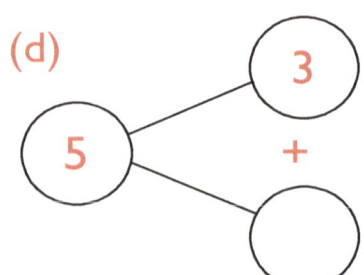

(e)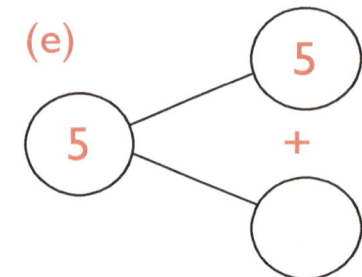

(f)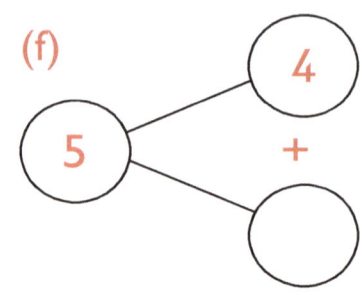

(g)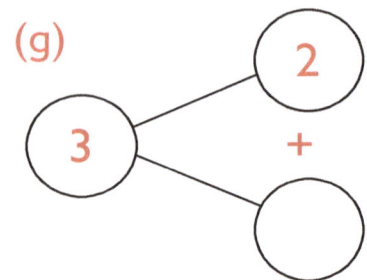

(h)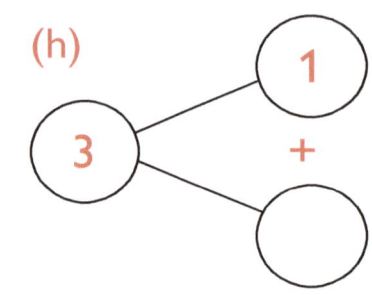

(i)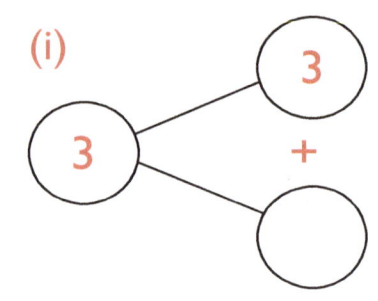

(j)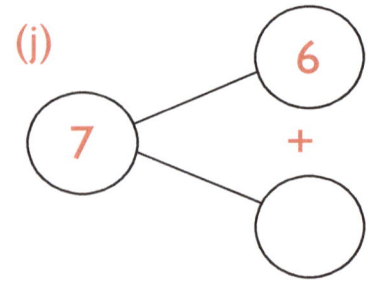

(k)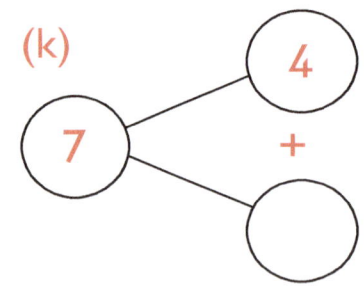

(l)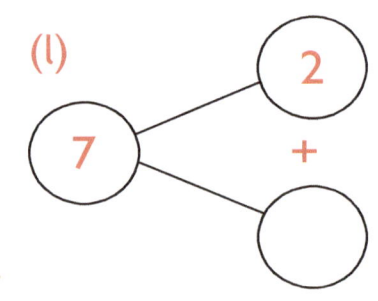

(m)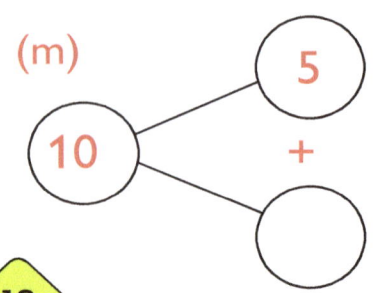

(n)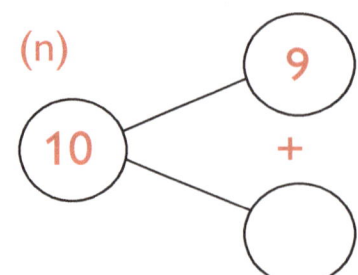

(o)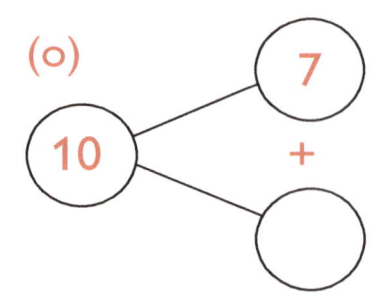

Subtraction Facts

1. Subtract the numbers below to complete the subtraction facts.

(a)
$5 - 0 =$ ☐
$5 - 1 =$ ☐
$5 - 2 =$ ☐
$5 - 3 =$ ☐
$5 - 4 =$ ☐
$5 - 5 =$ ☐

(b)
$6 - 0 =$ ☐
$6 - 1 =$ ☐
$6 - 2 =$ ☐
$6 - 3 =$ ☐
$6 - 4 =$ ☐
$6 - 5 =$ ☐
$6 - 6 =$ ☐

(c)
$4 - 0 =$ ☐
$4 - 1 =$ ☐
$4 - 2 =$ ☐
$4 - 3 =$ ☐
$4 - 4 =$ ☐

(d)
$3 - 0 =$ ☐
$3 - 1 =$ ☐
$3 - 2 =$ ☐
$3 - 3 =$ ☐

(e)
$1 - 0 =$ ☐
$1 - 1 =$ ☐

(f)
$2 - 0 =$ ☐
$2 - 1 =$ ☐
$2 - 2 =$ ☐

2. Complete the table given below using the subtraction facts.

−	9↓	8↓	7↓	6↓	5↓	4↓	3↓	2↓	1↓
1 →	8				4			1	0
2 →	7		5	4			1	0	
3 →		5			2		0		
4 →			3			0			
5 →	4			1	0				
6 →		2	1	0					
7 →			0						
8 →	1	0							
9 →	0								

3. Follow the given steps and fill in the blanks.

(a)

Take number 6
Add 2 = ☐
Subtract 5 = ☐
Add 3 = ☐
Subtract 1 = ☐
Answer is ☐

(b)

Take number 8
Subtract 7 = ☐
Add 3 = ☐
Subtract 3 = ☐
Add 8 = ☐
Answer is ☐

Shapes

1. **Join the dots, colour the shapes and name them.**

(a)

(b)

(c)

(d)

(e)

(f)

(g)

(h)

Star   Triangle   Square   Rectangle

Oval   Circle   Heart   Semicircle

2. Circle the objects which have the same shape.

3-D Shape	Objects
(a) Sphere	
(b) Cube	
(c) Cuboid	
(d) Cylinder	
(e) Cone	

3. Complete the tables given below.

Flat Shapes	Name of the shape	Number of sides	Number of corners
(a) ☐	_____	_____	_____
(b) △	_____	_____	_____
(c) ○	_____	_____	_____
(d) ▭	_____	_____	_____
(e) ◯	_____	_____	_____

Solid Shapes	Name of the shape	Number of faces	Number of edges	Number of vertices
(f)	_____	_____	_____	_____
(g)	_____	_____	_____	_____
(h)	_____	_____	_____	_____
(i)	_____	_____	_____	_____
(j)	_____	_____	_____	_____

Counting blocks

1. Count the blocks and write the number in the box.

(a)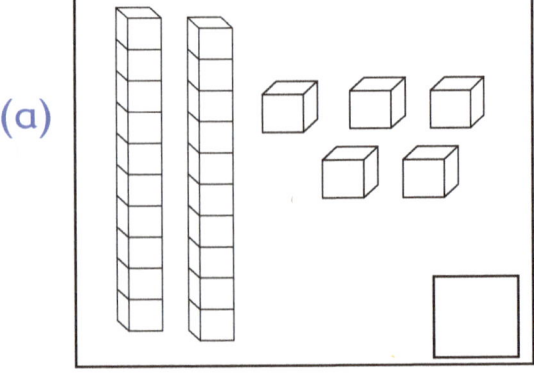

(b)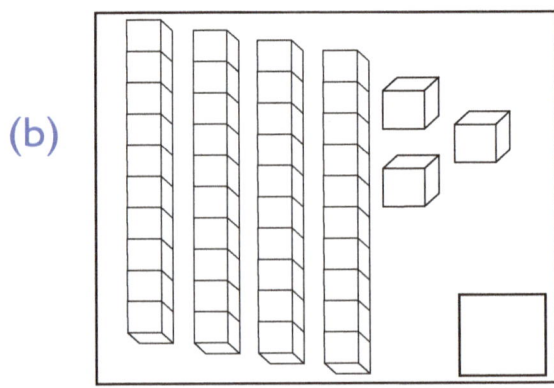

(c)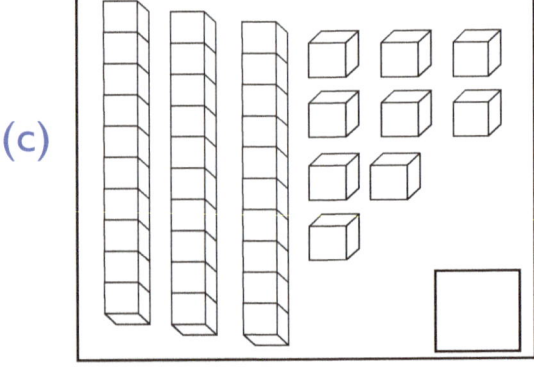

(d)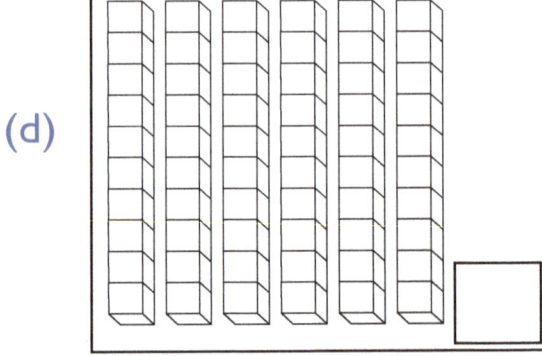

(e)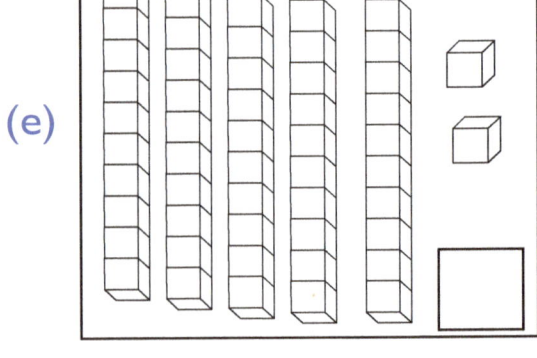

(f)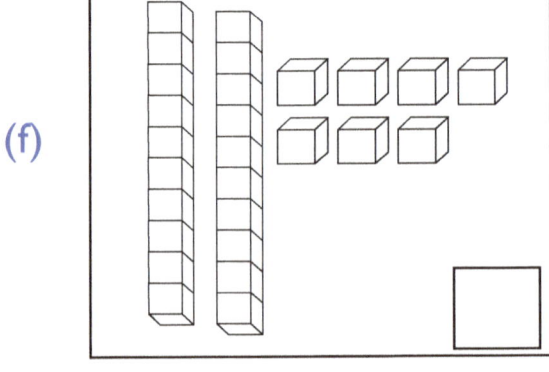

(g)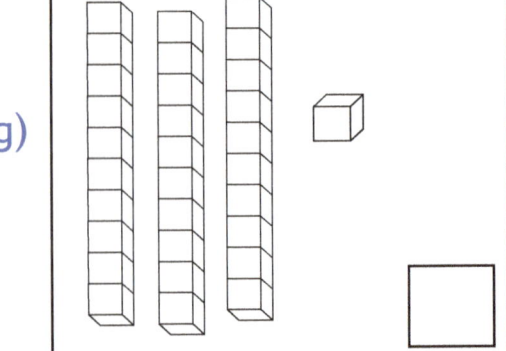

(h)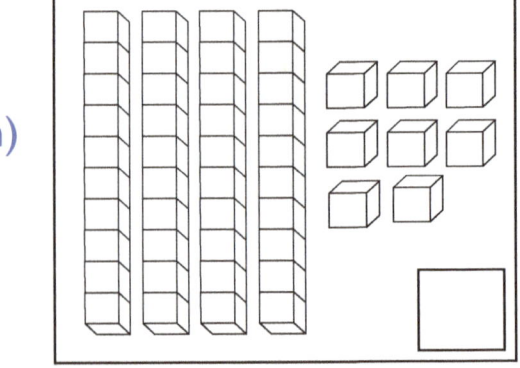

Tens and Ones

1. **Write the numbers in tens and ones and then write their number names.**

Number	Place T	Place O	Number name
(a) 53			
(b) 99			
(c) 87			
(d) 36			
(e) 61			

Number	Place T	Place O	Number name
(f) 28			
(g) 42			
(h) 15			
(i) 77			
(j) 50			

2. **Write the missing numbers upto 100.**

1						7			
		13						19	
21					26				30
			34					39	
	42						48		
		53			56				
61				65			68		
		74						79	
	82					87			
		93							100

Biggest and Smallest

1. **Tick the biggest number and cross out the smallest number in each group.**

(a) | 42 | 36 | 29 | 97 |

(b) | 55 | 68 | 85 | 56 |

(c) | 100 | 49 | 8 | 93 |

(d) | 28 | 39 | 76 | 9 |

(e) | 62 | 94 | 78 | 80 |

(f) | 15 | 66 | 84 | 32 |

(g) | 75 | 91 | 14 | 83 |

(h) | 48 | 49 | 46 | 40 |

2. **Write the smallest number in each group.**

(a) | 24 | 49 | 28 | |

(b) | 36 | 97 | 42 | |

(c) | 78 | 83 | 56 | |

(d) | 67 | 41 | 85 | |

(e) | 99 | 27 | 81 | |

(f) | 79 | 8 | 92 | |

3. **Write the biggest number in each group.**

(a) | 64 | 72 | 19 | |

(b) | 92 | 64 | 89 | |

(c) | 53 | 81 | 79 | |

(d) | 67 | 27 | 18 | |

(e) | 14 | 44 | 23 | |

(f) | 55 | 37 | 60 | |

Ordering Larger Numbers

1. Write the numbers in increasing order.

(a)

39	12	47	59	86

→

(b)

73	26	90	51	9

→

(c)

81	33	75	64	14

→

(d)

36	31	39	37	30

→

(e)

40	70	10	60	80

→

2. Write the numbers in decreasing order.

(a)

74	36	52	12	18

→

(b)

90	20	50	80	30

→

(c)

28	87	41	65	72

→

(d)

81	62	93	54	15

→

(e)

16	79	99	26	59

→

Number Names

1. Write the number names for the given numbers.

(a) 43 _____

(b) 62 _____

(c) 78 _____

(d) 19 _____

(e) 81 _____

(f) 57 _____

(g) 24 _____

(h) 36 _____

(i) 56 _____

(j) 88 _____

(k) 14 _____

(l) 49 _____

(m) 52 _____

(n) 75 _____

2. Read the number names and write the numbers.

(a) Sixty four ☐

(b) Ninety nine ☐

(c) Forty seven ☐

(d) Eighteen ☐

(e) Eighty three ☐

(f) Fifty one ☐

(g) Twenty two ☐

(h) Sixty eight ☐

(i) Thirty nine ☐

(j) Ninety four ☐

(k) Eighty five ☐

(l) Forty six ☐

(m) Eleven ☐

(n) Thirty five ☐

Adding Bigger Numbers

1. Regroup the numbers and add.

(a)

T	O
2	8
+ 3	2

(b)

T	O
5	5
+ 2	5

(c)

T	O
4	6
+ 1	6

(d)

T	O
3	7
+ 5	5

(e)

T	O
2	4
+ 6	6

(f)

T	O
7	7
+	5

(g)

T	O
5	6
+ 1	5

(h)

T	O
5	3
+ 2	8

(i)

T	O
4	9
+ 4	1

(j)

T	O
2	9
+ 3	2

(k)

T	O
5	7
+ 3	6

(l)

T	O
1	6
+ 1	5

(m)

T	O
2	8
+ 1	4

(n)

T	O
3	7
+ 4	5

(o)

T	O
1	3
+ 2	7

(p)

T	O
4	6
+ 2	6

(q)

T	O
6	2
+ 1	9

(r)

T	O
4	9
+ 3	6

(s)

T	O
7	8
+ 1	5

(t)

T	O
2	6
+ 5	6

Subtraction with Regrouping

1. Regroup the numbers and subtract.

(a)

T	O
7	3
− 2	8

(b)

T	O
5	0
− 2	7

(c)

T	O
6	0
− 5	4

(d)

T	O
3	3
− 2	9

(e)

T	O
9	0
− 4	6

(f)

T	O
8	7
− 4	8

(g)

T	O
9	1
− 2	5

(h)

T	O
5	7
− 3	9

(i)

T	O
5	4
− 2	6

(j)

T	O
8	0
− 2	2

(k)

T	O
7	6
− 4	7

(l)

T	O
2	3
− 1	4

(m)

T	O
8	3
− 5	1

(n)

T	O
6	4
− 2	2

(o)

T	O
3	9
− 2	0

(p)

T	O
5	2
− 3	1

(q)

T	O
6	8
− 2	5

(r)

T	O
4	9
− 1	8

(s)

T	O
3	7
− 3	0

(t)

T	O
7	9
− 5	9

Checking Answers

1. Solve the given problems. Rewrite the numbers in the grid as shown by the arrows and solve again to check your answer.

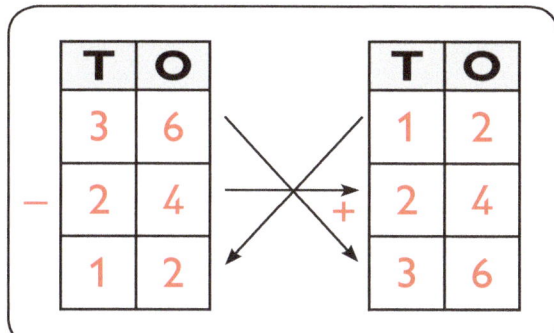

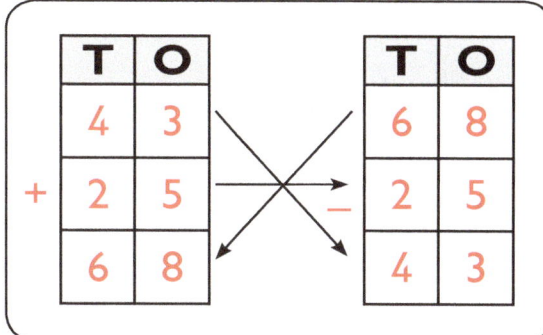

(a)

T	O
5	8
− 1	4

(b)

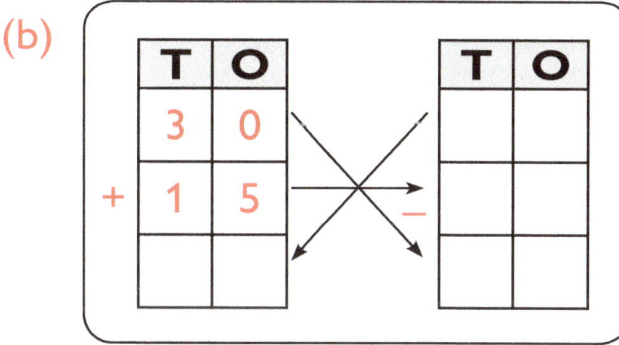

(c)

T	O
8	8
− 3	7

(d)

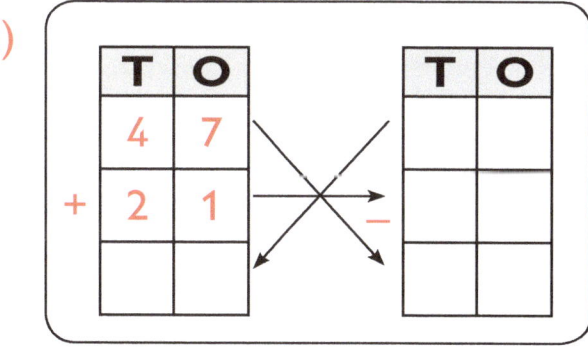

(e)

T	O
5	5
− 3	8

(f)

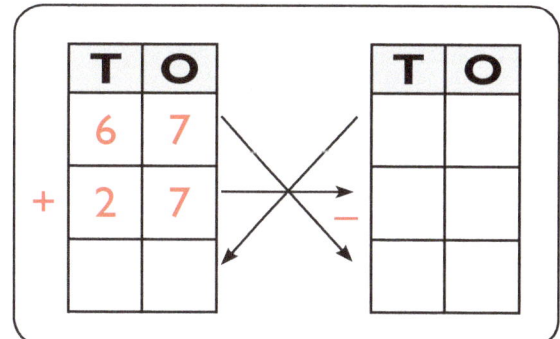

Word Problems

1. Read the word problems below and solve them.

T	O

(a) There are 21 red balls in a bag. Billy put 18 more balls in the same bag. How many balls are there in all? +

T	O

(b) Dolly makes 57 muffins to sell. She sells 29 muffins. How many muffins are left? −

T	O

(c) There are 49 flowers in a bunch. 30 flowers are pink and the others are yellow. How many yellow flowers are there in the bunch? −

T	O

(d) Juno has 35 marbles. He collected 48 more marbles. How many marbles does he have in all? +

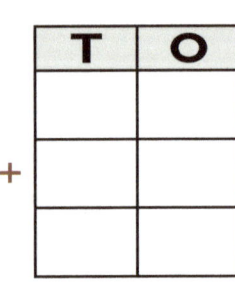

(e) There are 28 mangoes in one basket and 29 mangoes in another basket. How many mangoes are there in all? +

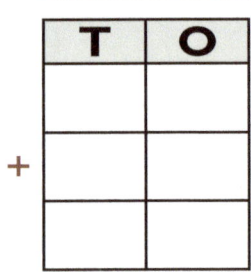

(f) Max has 52 candies. He gives 26 to his sister. How many candies are left with Max? −

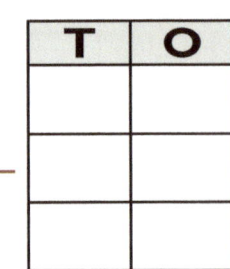

2. Use repeated addition to multiply the numbers.

(a) $2 + 2 + 2 + 2 + 2 + 2 =$ ☐ | $6 \times 2 =$ ☐

(b) $3 + 3 + 3 + 3 + 3 =$ ☐ | $5 \times 3 =$ ☐

(c) $4 + 4 + 4 + 4 + 4 + 4 =$ ☐ | $6 \times 4 =$ ☐

(d) $8 + 8 + 8 + 8 =$ ☐ | $4 \times 8 =$ ☐

(e) $6 + 6 + 6 + 6 + 6 + 6 + 6 =$ ☐ | $7 \times 6 =$ ☐

(f) $7 + 7 + 7 + 7 + 7 =$ ☐ | $5 \times 7 =$ ☐

(g) $2 + 2 + 2 + 2 + 2 + 2 + 2 + 2 =$ ☐ | $8 \times 2 =$ ☐

(h) $9 + 9 + 9 + 9 + 9 + 9 =$ ☐ | $6 \times 9 =$ ☐

(i) $5 + 5 + 5 + 5 + 5 + 5 + 5 =$ ☐ | $7 \times 5 =$ ☐

(j) $10 + 10 + 10 + 10 =$ ☐ | $4 \times 10 =$ ☐

33

Multiplication

1. **Count the number of objects given in the pictures below to complete the multiplication statement.**

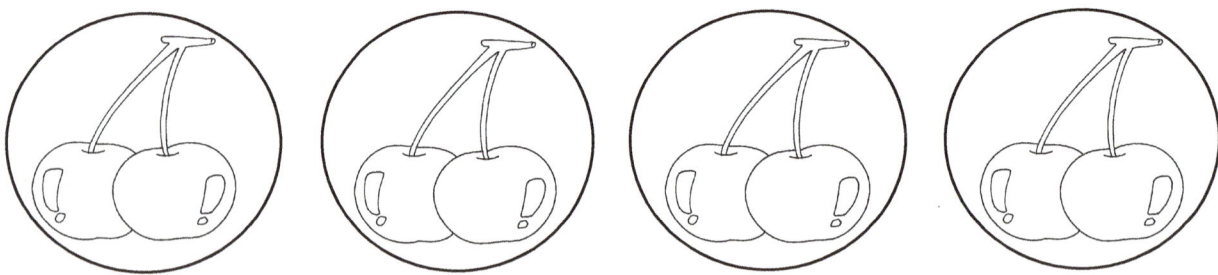

(a) There are _____ groups of cherries.

(b) There are _____ cherries in each group.

(c) There are _____ cherries in all.

2.

(a) There are _____ groups of candies.

(b) There are _____ candies in each group.

(c) There are _____ candies altogether.

3.

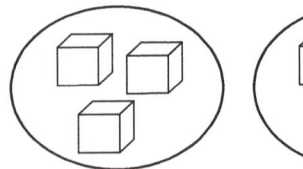 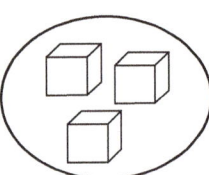 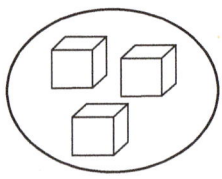 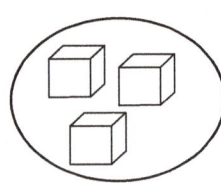 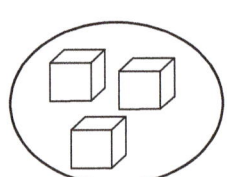

(a) There are _____ groups of blocks.

(b) There are _____ blocks in each group.

(c) There are _____ blocks in all.

4. Multiply the following.

(a)

T	O
	2
×	4

(b)

T	O
	3
×	3

(c)

T	O
1	4
×	2

(d)

T	O
2	3
×	2

(e)

T	O
3	5
×	2

(f)

T	O
1	2
×	3

(g)

T	O
1	5
×	3

(h)

T	O
2	9
×	2

(i)

T	O
1	8
×	5

(j)

T	O
2	6
×	3

(k)

T	O
1	4
×	3

(l)

T	O
3	6
×	2

(m)

T	O
1	3
×	4

(n)

T	O
2	9
×	2

(o)

T	O
4	2
×	2

(p)

T	O
3	0
×	3

(q)

T	O
4	4
×	2

(r)

T	O
2	0
×	4

(s)

T	O
1	6
×	3

(t)

T	O
1	9
×	5

Multiply and Colour

1. **Multiply the numbers on the balloons to find their product. Colour the balloons red that have even numbers as products and colour the others blue.**

(a)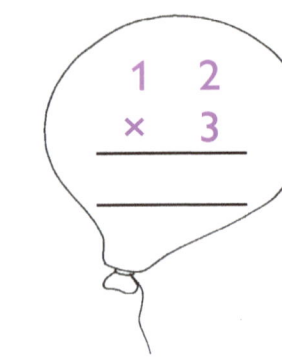
$$\begin{array}{r} 1\ 2 \\ \times\quad 3 \\ \hline \end{array}$$

(b)
$$\begin{array}{r} 1\ 5 \\ \times\quad 3 \\ \hline \end{array}$$

(c)
$$\begin{array}{r} 2\ 4 \\ \times\quad 3 \\ \hline \end{array}$$

(d)
$$\begin{array}{r} 1\ 4 \\ \times\quad 4 \\ \hline \end{array}$$

(e)
$$\begin{array}{r} 2\ 5 \\ \times\quad 2 \\ \hline \end{array}$$

(f)
$$\begin{array}{r} 1\ 3 \\ \times\quad 3 \\ \hline \end{array}$$

(g)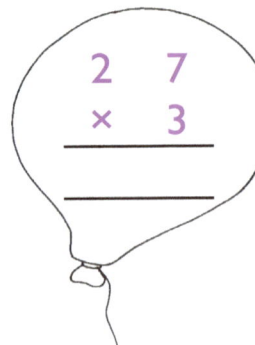
$$\begin{array}{r} 2\ 7 \\ \times\quad 3 \\ \hline \end{array}$$

(h)
$$\begin{array}{r} 2\ 9 \\ \times\quad 2 \\ \hline \end{array}$$

(i)
$$\begin{array}{r} 1\ 7 \\ \times\quad 4 \\ \hline \end{array}$$

(j)
$$\begin{array}{r} 1\ 2 \\ \times\quad 3 \\ \hline \end{array}$$

(k)
$$\begin{array}{r} 2\ 1 \\ \times\quad 3 \\ \hline \end{array}$$

(l)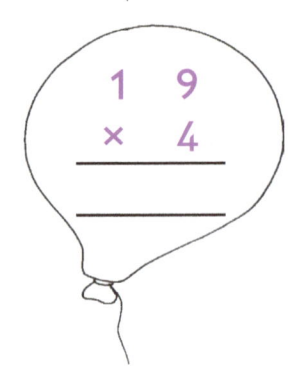
$$\begin{array}{r} 1\ 9 \\ \times\quad 4 \\ \hline \end{array}$$

Comparing Objects

1. Read and Circle.

(a) The object which is shorter than you.

Water Bottle	Pole

(b) The object which is taller than you.

Door	Pencil

(c) The object which is bigger than a cat.

Table	Eraser

(d) The object which is smaller than a dog.

Car	Rat

(e) The object which is heavier than your bag.

Bicycle	Comb

(f) The object which is lighter than a pen.

Lunch Box	Paper

(g) The object which is bigger than a cup.

Paper Clip	Chair

Measurement

1. **Measure the length of the given objects using given parts of your body.**

(a)   The length of your desk is _____ handspans.

(b)   The length of your classroom is _____ paces.

(c)   The length of your classroom is _____ footspans.

(d)   The width of the blackboard is _____ cubits.

(e)   The length of the window is _____ handspans.

(f)   The width of the door is _____ handspans.

(g)   The height of the teacher's table is _____ handspans.

(h)   Your height is equal to _____ handspans.

2. **Measure the length of the given objects using other objects and fill in the blanks.**

(a)   Your maths book is _____ paper pins long.

(b)   Your desk is _____ pencils long.

(c)   Your notebook is _____ sharpners long.

(d)   The blackboard is _____ pencils long.

(e)   The pencil case is _____ paper pins long.

(f)   Your pencil is _____ sharpners long.

Money

1. Fill in the blanks with the correct words.

head	tail	faces	Notes	Coins	metal

(a) _____ and notes are two common forms of money.

(b) Coins are flat pieces of _____ used as money.

(c) _____ are a form of paper money.

(d) There are two _____ of a coin.

(e) The two faces of a coin are called _____ and _____.

2. Read the restaurant menu below and answer the questions that follow.

Menu	
Items	**Price**
Burger	$18
Chips	$5
Coffee	$4
Nachos	$10
Cola drink	$6
Cupcake	$9
Fries	$14

(a) Jim had a burger, a coffee and a cupcake. How much money did he pay?

Answer _____

(b) Sara had fries, nachos, chips and a cola drink. How much money did she pay?

Answer _____

3. Solve the money problems given below.

(a) Tina has $66 in her purse. She bought a shirt for $28. How much money is she left with?

$\$\underline{\hspace{2cm}}$

(b) Mr Ben bought a can of juice for $18 and a bottle of milk for $25. How much money did he pay to the shopkeeper?

$\$\underline{\hspace{2cm}}$

(c) Harsh gave a $50 note to the shopkeeper. He bought a notebook for $15. How much money will he get back as change?

$\$\underline{\hspace{2cm}}$

(d) The price of an ice-cream is $12 and that of a pocorn is $18. If Meg buys both the items, how much will she spend?

$\$\underline{\hspace{2cm}}$

Days of the Week

1. **The names of the days of the week have been jumbled up. Unjumble them and rewrite the correct names.**

   (a) YNODMA   _____

   (b) ETDUAYS   _____

   (c) NWDESDEAY   _____

   (d) HSAYTURD   _____

   (e) YFARDI   _____

   (f) TUASRDYA   _____

   (g) NYUASD   _____

Months in a Year

1. Fill in the blanks with the names of the months.

(a) _____ is the first month of the year.

(b) _____ is the last month of the year.

(c) _____ comes after the month of July.

(d) _____ has the least number of days.

(e) _____ comes before the month of May.

(f) The month of _____ has the shortest name.

(g) The month of _____ has the longest name of all.

(h) _____ comes between June and August.

2. Write the number of days in the given months.

(a) January ☐   (g) March ☐

(b) February ☐   (h) December ☐

(c) July ☐   (i) September ☐

(d) June ☐   (j) April ☐

(e) May ☐   (k) October ☐

(f) August ☐   (l) November ☐

Time

1. **Look at the clocks given below. Write the time shown in each clock.**

(a) [_____]   (b) [_____]   (c) [_____]

(d) [_____]   (e) [_____]   (f) [_____]

2. **Draw the hands of the clocks to show the time at which you perform the given tasks. Also write the time in numbers.**

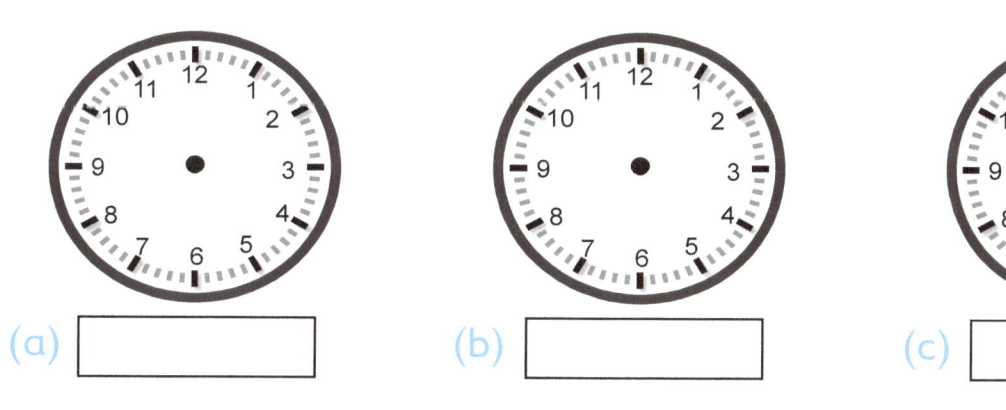

(a) [_____]   (b) [_____]   (c) [_____]

| Time at which you go to school. | Time at which you reach back home. | Time at which you go to sleep. |

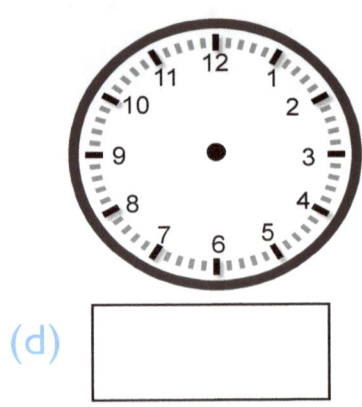

(d)

(e)

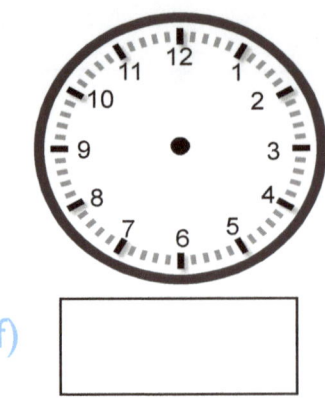

(f)

Time at which you wake up in the morning.	Time at which you have your lunch.	Time at which you have your dinner.

3. Draw the hands of the clocks to show the time given.

(a) 7:00

(b) 1:00

(c) 11:00

(d) 9:00

(e) 3:00

(f) 6:00

Word Search

1. **Search the number names from 1 to 10 in the given word search and write them below.**

U	I	T	E	N	I	H	E	M
X	C	F	I	V	E	L	L	N
S	I	X	T	S	E	V	E	N
S	O	N	E	A	N	I	N	E
T	H	R	E	E	F	A	T	A
C	V	T	L	F	O	U	R	O
T	T	W	O	E	I	G	H	T

1. _____   6. _____

2. _____   7. _____

3. _____   8. _____

4. _____   9. _____

5. _____   10. _____

Answers

Count the Objects

1. (a) 5   (b) 10   (c) 7
    (d) 8   (e) 16   (f) 3

2. (a) 3, 4, 6, 8
    (b) 9, 10, 11, 13
    (c) 14, 16, 18, 19
    (d) 1, 2, 4, 5, 7
    (e) 8, 9, 11, 12, 14

Forward and Backward

1. (a) 2, 3, 4, 6, 7 8
    (b) 5, 6, 7, 9, 10, 11
    (c) 11, 12, 14, 15, 16, 18
    (d) 21, 22, 23, 24, 26, 28

2. (a) 9, 8, 7, 5, 4 , 2
    (b) 14, 13, 11, 10, 8, 7
    (c) 17, 16, 15, 13, 12, 10
    (d) 28, 26, 24, 23, 22, 21

Before, After and Between Numbers

1. (a) 5   (b) 8   (c) 12
    (d) 9   (e) 6   (f) 4
    (g) 1   (h) 10   (i) 2
    (j) 11

2. (a) 6   (b) 8   (c) 19
    (d) 10   (e) 2   (f) 12
    (g) 7   (h) 4   (i) 20
    (j) 15   (k) 11   (l) 13

3. (a) 16   (b) 4   (c) 9
    (d) 13   (e) 10   (f) 7
    (g) 2   (h) 15

Comparing Numbers

2. (a) >   (b) =   (c) =
    (d) >   (e) >   (f) <
    (g) <   (h) <   (i) >
    (i) =

Ordering Numbers

1. (a) 2, 3, 5, 7   (b) 1, 4, 6, 9
    (c) 1, 3, 8, 10   (d) 2, 4, 7, 9

2. (a) 8, 5, 4, 0   (b) 10, 6, 3, 2
    (c) 9, 7, 4, 1   (d) 10, 8, 5, 3

Addition up to 10

1. (a) $5 + 4 = 9$ (b) $4 + 2 = 6$
    (c) $7 + 0 = 7$

2. (a) 7   (b) 10   (c) 4
    (d) 5   (e) 8   (f) 10
    (g) 9   (h) 8   (i) 6

Add on the Number Line

1. (a) 7   (b) 4   (c) 9
    (d) 10   (e) 6   (f) 10

Subtraction up to 10

1. (a) 5   (b) 5   (c) 4
    (d) 3   (e) 4   (f) 6
    (g) 8

2. (a) 8   (b) 7   (c) 1
    (d) 6   (e) 2   (f) 4
    (g) 3   (h) 9   (i) 5

3. (a) 8   (b) 5   (c) 1
    (d) 3   (e) 9   (f) 7

4. (a) 5   (b) 3   (c) 7
    (d) 7   (e) 2   (f) 2
    (g) 6   (h) 0   (i) 0
    (j) 1   (k) 2   (l) 3

Subtract on the Number Line

1. (a) 5   (b) 0   (c) 2
    (d) 2   (e) 3   (f) 6

Addition Facts

2. (a) 4   (b) 7   (c) 1
    (d) 4   (e) 3   (f) 1
    (g) 2   (h) 0   (i) 6

Number Bonds

1. (a) 3   (b) 4   (c) 1
    (d) 2   (e) 0   (f) 1
    (g) 1   (h) 2   (i) 0
    (j) 1   (k) 3   (l) 5
    (m) 5   (n) 1   (o) 3

Subtraction Facts

1. (a) 5, 4, 3, 2, 1, 0
    (b) 6, 5, 4, 3, 2, 1, 0
    (c) 4, 3, 2, 1, 0
    (d) 3, 2, 1, 0
    (e) 1, 0
    (f) 2, 1, 0

3. (a) 5   (b) 9

Shapes

1. (a) Triangle   (b) Square
    (c) Circle   (d) Rectangle
    (e) Star   (f) Semicircle
    (g) Oval   (h) Heart

3. (a) Square, 4, 4
    (b) Triangle, 3, 3
    (c) Circle, 0, 0
    (d) Rectangle, 4, 4
    (e) Oval, 0, 0
    (f) Sphere, 1, 0, 0
    (g) Cube, 6, 12, 8
    (h) Cone, 2, 1, 1
    (i) Cuboid, 6, 12, 8
    (j) Cylinder 3, 2, 0

Counting Blocks

1. (a) 25   (b) 43   (c) 39
    (d) 60   (e) 52   (f) 27
    (g) 31   (h) 48

Biggest and Smallest

1. (a) B-97, S-29   (b) B-85, S-55
    (c) B-100, S-8   (d) B-76, S-9
    (c) B-94, S-62   (d) B-84, S-15
    (c) B-91, S-14   (d) B-49, S-40

2. (a) 24   (b) 36   (c) 56
    (d) 41   (e) 27   (f) 8
3. (a) 72   (b) 92   (c) 81
    (d) 67   (e) 44   (f) 60

Ordering Larger Numbers

1. (a) 12, 39, 47, 59, 86
    (b) 9, 26, 51, 73, 90
    (c) 14, 33, 64, 75, 81
    (d) 30, 31, 36, 37, 39
    (e) 10, 40, 60, 70, 80

2. (a) 74, 52, 36, 18, 12
    (b) 90, 80, 50, 30, 20
    (c) 87, 72, 65, 41, 28
    (d) 93, 81, 62, 54, 15
    (e) 99, 79, 59, 26, 16

Number Names

1. (a) Forty three
    (b) Sixty two
    (c) Seventy eight
    (d) Nineteen
    (e) Eighty one
    (f) Fifty seven
    (g) Twenty four
    (h) Thirty six
    (i) Fifty six
    (j) Eighty eight
    (k) Fourteen
    (l) Forty nine
    (m) Fifty two
    (n) Seventy five

2. (a) 64   (b) 99   (c) 47
    (d) 18   (e) 83   (f) 51
    (g) 22   (h) 68   (i) 39
    (j) 94   (k) 85   (l) 46
    (m) 11   (n) 35

Adding Bigger Numbers

1. (a) 60   (b) 80   (c) 62
    (d) 92   (e) 90   (f) 82
    (g) 71   (h) 81   (i) 90
    (j) 61   (k) 93   (l) 31
    (m) 42   (n) 82   (o) 40

(p) 72   (q) 81   (r) 85
(s) 93   (t) 82

Subtraction with Regrouping

1. (a) 45   (b) 23   (c) 6
    (d) 4   (e) 44   (f) 39
    (g) 66   (h) 18   (i) 28
    (j) 58   (k) 29   (l) 9
    (m) 32   (n) 42   (o) 19
    (p) 21   (q) 43   (r) 31
    (s) 7   (t) 20

Checking Answers

1. (a) 44   (b) 45   (c) 51
    (d) 68   (e) 17   (f) 94

Word Problems

1. (a) 39   (b) 28   (c) 19
    (d) 83   (e) 57   (f) 26

2. (a) 12   (b) 15   (c) 24
    (d) 32   (e) 42   (f) 35
    (g) 16   (h) 54   (i) 35
    (j) 40

Multiplication

1. (a) 4   (b) 2   (c) 8
2. (a) 3   (b) 4   (c) 12
3. (a) 5   (b) 3   (c) 15
4. (a) 8   (b) 9   (c) 28
    (d) 46   (e) 70   (f) 36
    (g) 45   (h) 58   (i) 90
    (j) 78   (k) 42   (l) 72
    (m) 52   (n) 58   (o) 84
    (p) 90   (q) 88   (r) 80
    (s) 48   (t) 95

Multiply and Colour

1. (a) 36   (b) 45   (c) 72
    (d) 56   (e) 50   (f) 39
    (g) 81   (h) 58   (i) 68
    (j) 36   (k) 63   (l) 76

Comparing Objects

1. (a) Water bottle (b) Door (c) Table
    (d) Rat   (e) Bicycle (f) Paper
    (g) Chair

Money

1. (a) Coins   (b)   metal
    (c) Notes   (d)   faces
    (e) head, tail

2. (a) $31   (b) $35

3. (a) $38   (b) $43
    (c) $35   (d) $30

Days of the Week

1. (a) Monday   (b) Tuesday
    (c) Wednesday   (d) Thursday
    (e) Friday   (f) Saturday
    (g) Sunday

Months in a Year

1. (a) January   (b) December
    (c) August   (d) February
    (e) April   (f) May
    (g) September   (h) July

2. (a) 31   (b) 28 or 29   (c) 31
    (d) 30   (e) 31   (f) 31
    (g) 31   (h) 31   (i) 30
    (j) 30   (k) 31   (l) 30

Time

1. (a) 2:00   (b) 5:00
    (c) 11:00   (d) 12:00
    (e) 3:00   (f) 10:00